EARLY RED JACKET AND CALUMET IN PICTURES

VOL. II

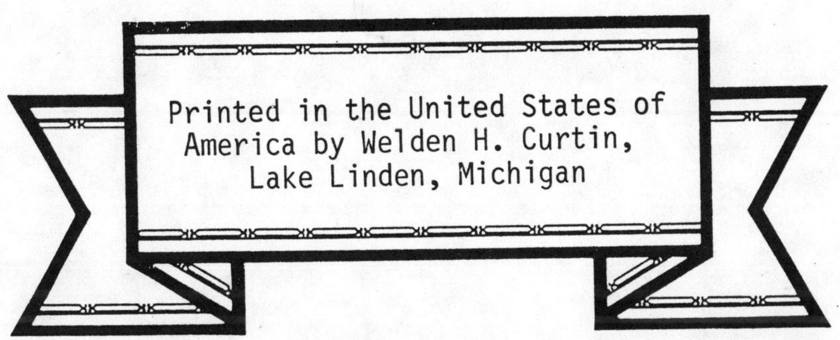

COVER PICTURE: This photo, "A Fourth
of July Parade in Calumet" was taken
about the turn of the century by A. F.
Isler, a popular Copper Country
commercial photographer. Courtesy of
David Mac Frimodig of Laurium,
Michigan.

AUTHOR'S NOTE

This second publication was complied to further highlight Red Jacket's and Calumet's early history. As mentioned before, it is very difficult to describe in words what this village looked like in the "olden days" and the undersigned hopes that the selection and publication of these turn of the century pictures will help. Many local citizens are interested in their early heritage and especially so as the National Park initiative moves forward.

All of these photographs reproduced are from private collections. On behalf of the Copper Country historians, I wish to thank them for their generosity and support.

Clarence J. Monette
Author

Building a smoke stack at the
Calumet and Hecla Mining Company

Courtesy of Anthony L. Bausano, Copper World, Calumet, Michigan

This picture of the College Inn, located on Fifth Street, was taken in 1920

Courtesy of Anthony L. Bausano, Copper World, Calumet, Michigan

Blind workers employed at the C & H broom factory in Calumet. Courtesy of the David "Mac" Frimodig collection.

Blind workers employed at the C & H broom factory in Calumet. Courtesy of the David "Mac" Frimodig collection.

The Salvation Army Barracks as seen in about 1899

Salvation Army Band (Calumet) 1925
Top Row: Thursa Rule Harvey, Emily Matthews, Bill Barnes, Howard Harvey, Lillian Benny Strom, Ethel Bloy Lager.
Second Row: Vernon Strom, Buster Mitchell, Dorothy Strom Kumpala, Bill Matthews, Peggy Guinnen, Hank Lucas, Theodore Strom.
First Row: Clarice Lucas Burton, Rudy Strom, Mrs. Capt. Agre, Capt. Agre, Francis Benny, Harry Lloyd, William Harvey.
Seated at Front: Carl Thelin, Marshall Benny.

This picture of the C&H Hancock and Pewabic hoist was taken from the Swedetown hill. Courtesy of the David "Mac" Framodig collection.

The first Washington and Jefferson school was located on Calumet Avenue. This picture was taken in 1900 and is through the courtesy of the David "Mac" Framodig collection.

The old boilerhouse and stacks from the
Calumet and Hecla Stamp Mill

GATELY-WIGGINS COMPANY
Courtesy of Anthony L. Bausano, Copper
World, Calumet, Michigan

Sacred Heart Church, Larium, Mich.
Sacred Heart Church, Larium, Mich.

The Sacred Heart Church was destroyed
by fire on August 13, 1983 after 86
years of service

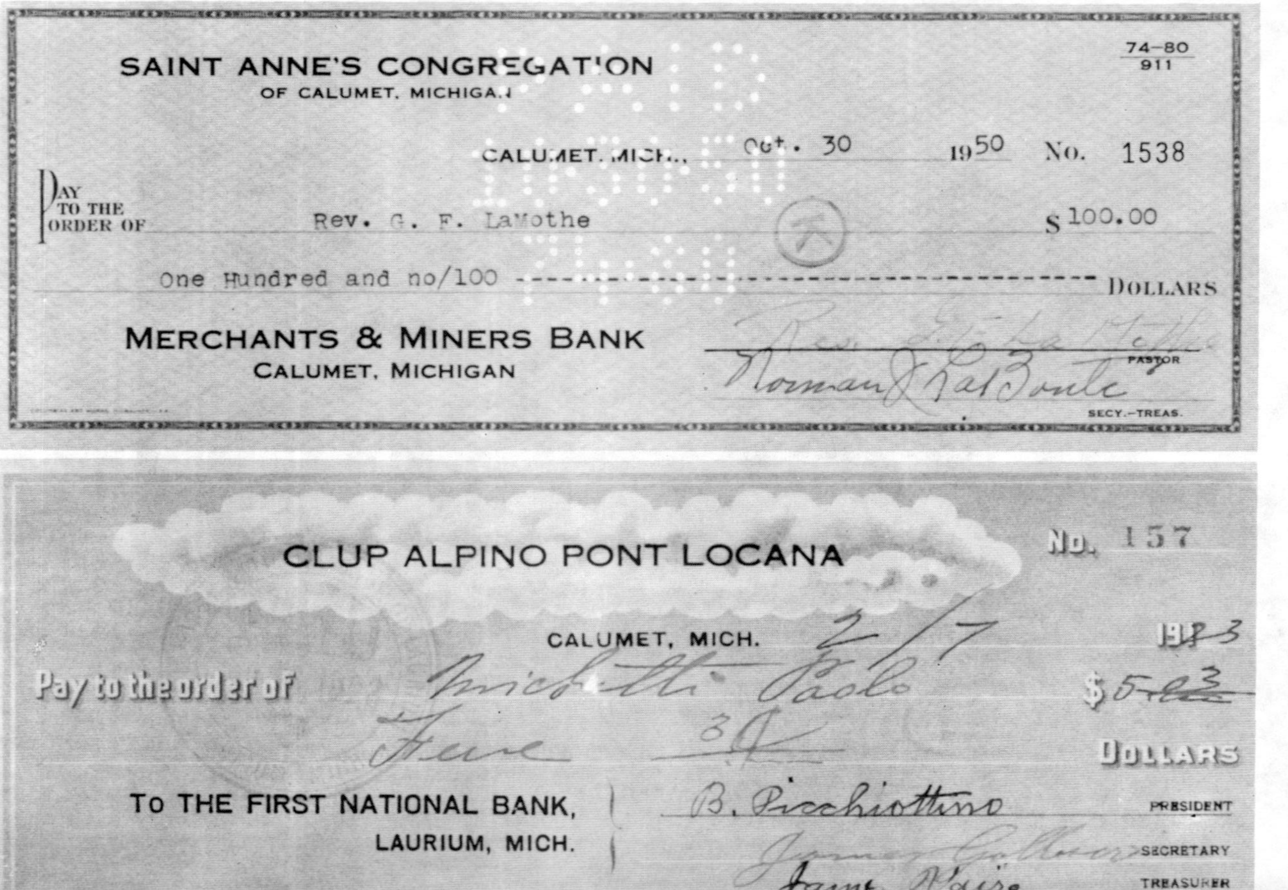

SAINT ANNE'S CONGREGATION
OF CALUMET, MICHIGAN
74-80
911
CALUMET, MICH., Oct. 30 1950 No. 1538
PAY TO THE ORDER OF Rev. G. F. LaMothe $100.00
One Hundred and no/100 ------------------------------ DOLLARS
MERCHANTS & MINERS BANK
CALUMET, MICHIGAN
Rev. G. F. LaMothe
PASTOR
Norman J. LaBonte
SECY.-TREAS.

CLUP ALPINO PONT LOCANA
No. 157
CALUMET, MICH. 2/7 1923
Pay to the order of Nicoletti Paolo $5.23
Five 3/ DOLLARS
TO THE FIRST NATIONAL BANK,
LAURIUM, MICH.
B. Picchiottino
PRESIDENT
SECRETARY
TREASURER

CERTIFICATE FOR LESS THAN ONE HUNDRED SHARES
CERTIFICATE FOR LESS THAN ONE HUNDRED SHARES
SHARES $25, EACH.
SHARES $25, EACH.
NUMBER
B0985
SHARES
++9++
INCORPORATED UNDER THE LAWS   FULL PAID AND NON-ASSESSABLE   OF THE STATE OF MICHIGAN,
CALUMET AND HECLA CONSOLIDATED COPPER COMPANY
THIS CERTIFICATE IS TRANSFERABLE EITHER IN NEW YORK OR BOSTON
This Certifies that   FLORENCE E. GRANT+ + + + + +
is the owner of   + + + + NINE + + + +   full paid and non
assessable shares of the Capital Stock of the Calumet and Hecla Consolidated Copper Company
transferable on the books of the Company by the holder hereof in person or by attorney on the
surrender of this certificate properly endorsed. This certificate is not valid until countersigned
by the Transfer Agent and Registered by the Registrar.
Witness, the seal of the Company   and the signatures of its duly authorized
officers this   DEC 21 1923
REGISTERED:
AMERICAN TRUST COMPANY
(BOSTON)
By
ASSISTANT TREASURER
COUNTERSIGNED:
OLD COLONY TRUST COMPANY
(BOSTON)
By
ASSISTANT SECRETARY
ASSISTANT TO THE PRESIDENT

Calumet Post Office

OFFICE OF THE POSTMASTER.

CALUMET, HOUGHTON CO., MICH., *Mar 23rd* 1905

DR. SORSEN'S PRIVATE HOSPITAL.

C. J. SORSEN, M. D.,
SURGEON IN CHIEF

Calumet, Mich. *Feby 6th*

Village of Red Jacket

Houghton County   **Calumet, Mich.**

June 17, 1907.

M.F.Foley,County Treas.,

  Houghton,Mich.

Dear Sir:--

  Enclosed please find liquor dealers'bond of

Jos.Schlitz Brewing Co.,

   Yours truly,

         Clerk.

Sixth Street, Calumet, Mich.

Courtesy of Anthony L. Bausano, Copper World, Calumet, Michigan

D. S. S. & A. R. R. STATION, IN WINTER, CALUMET, MICH.

Fifth Street, Calumet, Mich.

Calumet - looking north from the Number 14 engine house. Courtesy of the David "Mac" Framodig collection

The Tamarack School, Calumet, Michigan
Courtesy of Mrs Margaret Reilly, Laurium, Michigan

This double deck cage and skip was located in the Red Jacket mine shaft. Courtesy of the David "Mac" Framodig collection.

This picture of the Bethlehem Church on
Calumet's Agent Street was taken in
1910

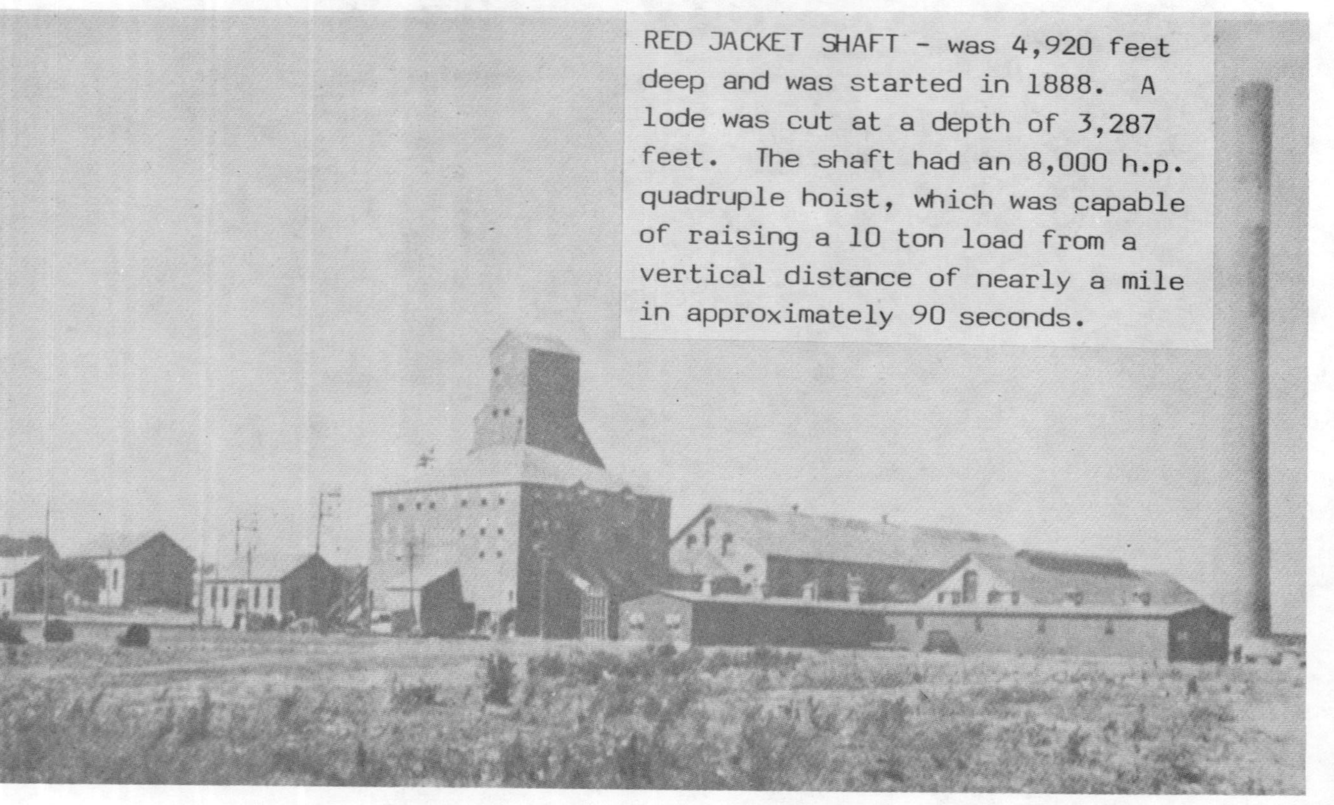

RED JACKET SHAFT - was 4,920 feet deep and was started in 1888. A lode was cut at a depth of 3,287 feet. The shaft had an 8,000 h.p. quadruple hoist, which was capable of raising a 10 ton load from a vertical distance of nearly a mile in approximately 90 seconds.

The Bethlehem church was located on
Agent Street, next to the C&H railroad
yeard. Courtesy of the David "Mac"
Framodig collection

CALUMET HIGH SCHOOL BASEBALL TEAM
CHAMPION INTERSCHOLASTIC LEAGUE 1913
Mothersill, Coach.
Savini, C.
Tornquist, S. S.
Horkins, L. F.
James, P.
Opland, 1st B.
Bennetts, 2nd B. Capt.
Tremberth, P.
Larson, C. F.
Nicholls, R. F.
Wier, P.
Anderson, H.
Gray, Pres. Athletic Association
McDuff, 3rd B.
Photo by J. W. Nara.

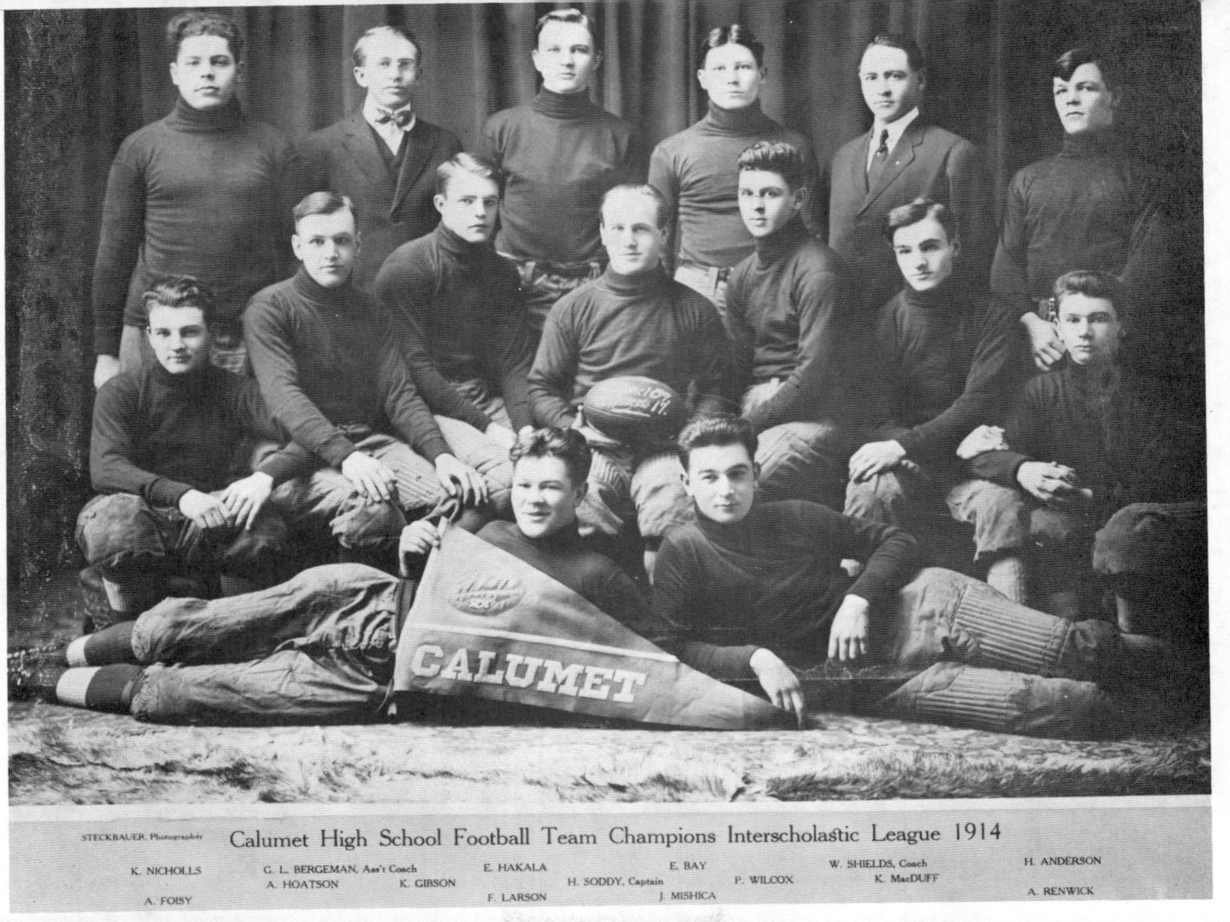

STECKBAUER, Photographer

Calumet High School Football Team Champions Interscholastic League 1914

K. NICHOLLS   G. L. BERGEMAN, Ass't Coach   E. HAKALA   E. BAY   W. SHIELDS, Coach   H. ANDERSON
A. HOATSON   K. GIBSON   H. SODDY, Captain   P. WILCOX   K. MacDUFF
A. FOISY   F. LARSON   J. MISHICA   A. RENWICK

One of Calumet and Hecla's surface work crews in Calumet

A Calumet baseball team in 1911

Rowe's Transportation Lines first new
moving van was delivered in 1941.
Courtesy of the Rowe Moving & Storage.

This picture of a Rowe Transportation
Line truck was taken in about 1928.
Courtesy of the Rowe Moving & Storage.

Several of Calumet's citizens on their
Indian motorcycles

An early photo of Rowe's Transportation Line fleet of trucks. Courtesy of the Rowe Moving and Storage Company.

1915 Natural Wall, Calumet, Mich.

Electric Park was located near Calumet and was visited by traveling on the Houghton County streetcar.

Taken on May 30, 1894, the Calumet
Light Guard plays near the Calumet Dam.
Courtesy of the David "Mac" Framodig
collection

Calumet as seen in 1900

BEN BLUM
LIQUOR STORE
CALUMET CORNER 5th. & PINE ~18
This picture was taken at the corner of
Fifth and Pine Streets in 1895

One of Calumet's stores

Part of the Calumet and Hecla Mining
Company complex

Underground in the Red Jacket Shaft

Calumet, Michigan

Just "plowing out"

_______ Mich, ______ Sep 1 190__

M Tri Mountain Mining Co

BOUGHT OF

BAJARI & ULSETH,
LUMBER YARD.
TELEPHONE CONNECTIONS.

Date	Qty		Description	Rate	Amount
Sep 1	5200	ft	Dressed Pine	16 00	83 20
	7443	"	Ship Lap	"	119 08
	638	"	Siding	21 00	13 40
	373	"	Flooring	28 00	9 05
	3150	"	Lath	2 50	7 88
	14 M		Shingles	2 60	36 40
	626	Lin ft	½ Round	50	3 13
	898	"	Base Mldg.	90	8 08
	868	"	Wnd Stops	45	3 90
	64	"	Door "	65	42
	130	"	O G Rail	45	59
	1	"	2 × 6 – 14 – 14 1/4 4931	14 50	33 75
	31	"	2 × 6 – 18 – 558	15 00	8 37
	36	"	2 × 8 – 12 – 576	14 50	8 35
	2	"	2 × 8 – 14 – 37	"	54
	18	"	2 × 8 – 22 – 578	16 00	8 71
	10	Pc	Wnd Casing O.S.	10	1 00
	242	"	" " I.S.	10	24 20
	44	"	" Jambs	25	11 00
	31	"	" Casing with Cap	5	1 55
	73	Pc	Stair Steps	10	2 30
	45	"	Window Sill	8	3 60
	2		Windows Small	80	1 60
	2	"	" No Glass		00
	2	"	" Large	1 50	3 00
	2		Doors	1 70	3 40
					396

Prices OK
Original

Goods Received G.
Audited A.
Paid 9/2 ?
Cash Folio 73

BARRETT'S
BIRD STORE.

A bird's eye view of Calumet

Dedication of the Soldiers Monument at
the Lakeview Cemetary in May of 1900

Mother Jones, one of Calumet's leaders.
Courtesy of the David "Mac" Frimodig
collection

This Calumet and Hecla Mining Company
Gear House was located on Mine Street.
Picture taken in 1886. Courtesy of the
David "Mac" Framodig collection.

GREENLEE PRINTING CO.

This store was located on
Calumet's Fifth Street

Courtesy of Anthony L. Bausano, Copper World, Calumet, Michigan

Map of the Calumet-Laurium area.

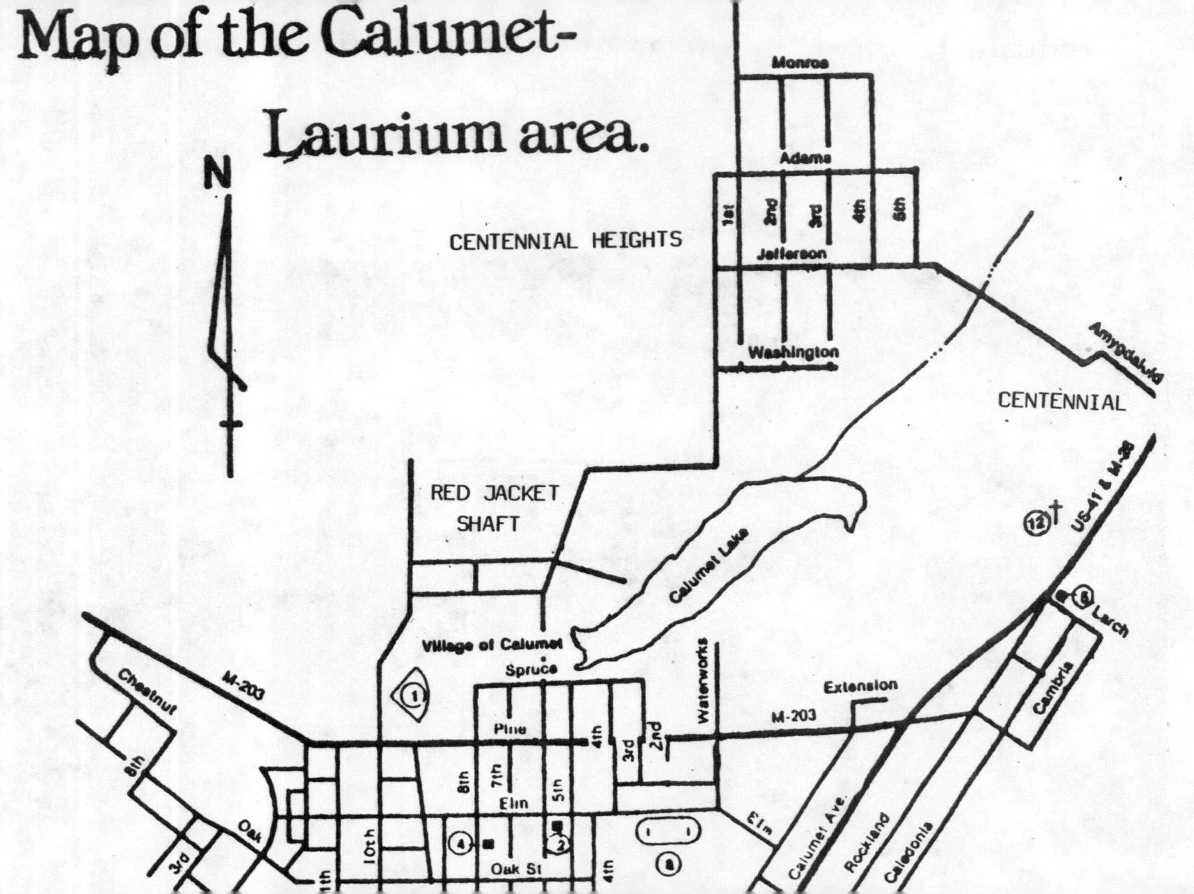

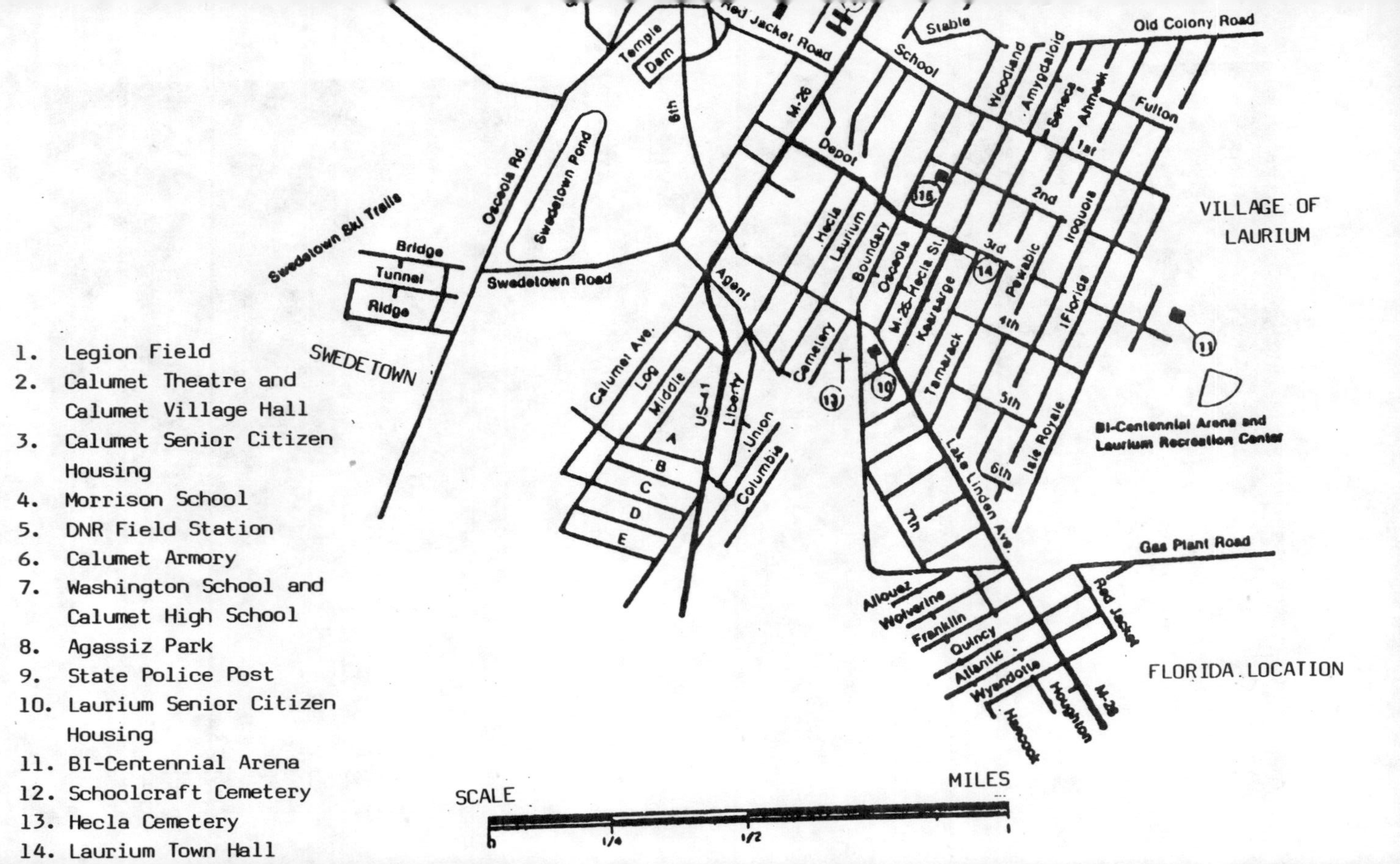

1. Legion Field
2. Calumet Theatre and
   Calumet Village Hall
3. Calumet Senior Citizen
   Housing
4. Morrison School
5. DNR Field Station
6. Calumet Armory
7. Washington School and
   Calumet High School
8. Agassiz Park
9. State Police Post
10. Laurium Senior Citizen
    Housing
11. BI-Centennial Arena
12. Schoolcraft Cemetery
13. Hecla Cemetery
14. Laurium Town Hall

The copper miner's
strike of 1913

Michigan National Guard encampment
during the 1913 strike

Michigan National Guard during
the 1913 copper miner's strike

The copper miner's
strike of 1913

Michigan National Guard during
the 1913 copper miner's strike

Michigan National Guard during the 1913 copper miner's strike

Michigan National Guard during the 1913 copper miner's strike

Michigan National Guard during
the 1913 copper miner´s strike

Michigan National Guard working
in the mine during the 1913 strike

The copper miner's strike of 1913

The copper miner's
strike of 1913

The copper miner's strike of 1913

The copper miner's
strike of 1913

Funeral victims in the
Italian Hall disaster, 1913

Calumet, Mich., June 1st _______ 1904.

Mr. Frank Schroeder.

TO JAMES McCLURE, DR.

Horses, Buggies, Wagons, Harness, Saddles, Etc.

BOUGHT. SOLD OR EXCHANGED.

BARN ON NORTH SEVENTH STREET.

| Apr. | 17 | Single. | | $2 | 00 | |

July 5 1904 Paid
James McClure

Carlton Hardware Company.

Agents for Worthington's Steam Pumps and Repairs.

Calumet, Mich., Feby. 15th, 1902.

Orders by mail or telephone promptly attended to.

Sold to Trimountain Mining Company,

Folio 117J127   Terms 30 Days Net.   Trimountain, Mich.

√2	Kegs 1½" x 1½" x 25/32" Sq Chuck Nuts		
√2	" 1½" x 1½" x 7/8" 800 lbs	4½ ✔	$ 36 00

F.O.B. Detroit, Mich.

4 Kegs Bolts & Nuts 820 lbs.

TELEPHONE 25

COR. SIXTH AND PORTLAND STS.

CALUMET, MICH., _______________ June 1 _____ 192 1

M_ Joseph Schroder _______________

AUTOMOBILES IN CONNECTION

TO **PIONEER LIVERY STABLE**, DR.
JOHN R. RYAN, PROPRIETOR
FUNERAL DIRECTOR

Oct 2ᵈ Team hauling dirt 9 AM to 5 PM 5⁰⁰

By check # 765

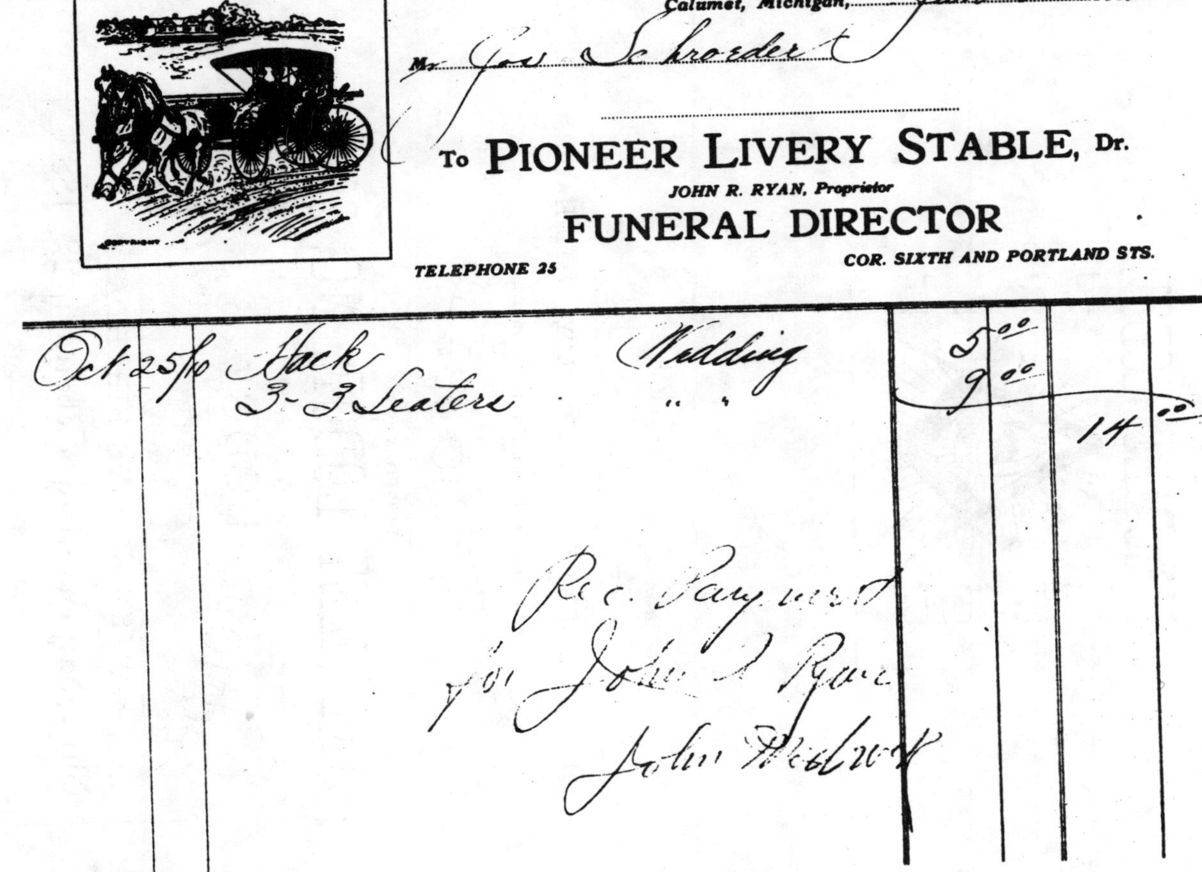

Calumet, Michigan, Jan 1 1918
Mr. Jos Schroeder
To PIONEER LIVERY STABLE, Dr.
JOHN R. RYAN, Proprietor
FUNERAL DIRECTOR
TELEPHONE 25
COR. SIXTH AND PORTLAND STS.
Oct 25/16 To Hack   Wedding   5 00
           3 - 3 Seaters   " "   9 00
                                                14 00
Rec. Payment
for John R Ryan
John Westwork

CALUMET LODGE,
NO. 271,
F. & A. M.

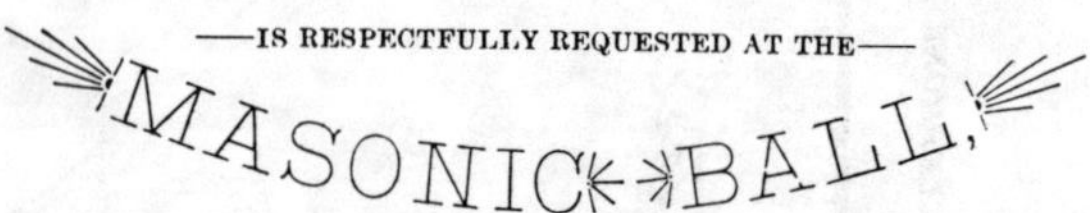

THE PLEASURE OF YOUR COMPANY, WITH LADIES,
—IS RESPECTFULLY REQUESTED AT THE—
MASONIC BALL,
—TO BE GIVEN BY THE—
CALUMET LODGE, NO. 271,
F. & A. M.
SCHOOL HOUSE HALL,
—ON—
WASHINGTON'S BIRTHDAY,
Thursday Evening, February 22, 1883.
DANCING AT 8 O'CLOCK.

Courtesy of Anthony L. Bausano, Copper World, Calumet, Michigan

Courtesy of Anthony L. Bausano, Copper World, Calumet, Michigan

The Washington School was first
occupied in 1875 and was
built by C & H Mining Co.
Enrollment was 6,327 in
1910-11 and had 225
teachers in 1915

Courtesy of Anthony L. Bausano, Copper World, Calumet, Michigan

Courtesy of Anthony L. Bausano, Copper World, Calumet, Michigan

The 1888 Number 3 Shaft Fire
Calumet and Hecla Mining Company
Courtesy of Mac Frimodig, Calumet, MI

The 1884 Hecla Mine Fire
Calumet and Hecla Mining Company
Courtesy of Mac Frimodig, Calumet, MI

Courtesy of Anthony L. Bausano, Copper World, Calumet, Michigan

Courtesy of Anthony L. Bausano, Copper World, Calumet, Michigan

Mineral Range Railroad
Depot, Calumet
Michigan

The Glass Block Store was destroyed by
fire on May 17, 1917.

Courtesy of Anthony L. Bausano,
Copper World, Calumet, Michigan

An early view of Calumet
showing the Tamarack Location

November 1, 1980

Destroyed by fire on
December 19, 1990

The children's reading room of the Calumet and Hecla Mining Company
Library, taken in 1906. Courtesy of David Mac Frimodig, Laurium, Mich

The Calumet and Hecla Mining Company Library, Calumet, Michigan

Courtesy of Anthony L. Bausano, Copper World, Calumet, Michigan

This building was constructed between 1880 and 1890 and
is now the D&N Branch Office at 330 Fifth Street, Calumet

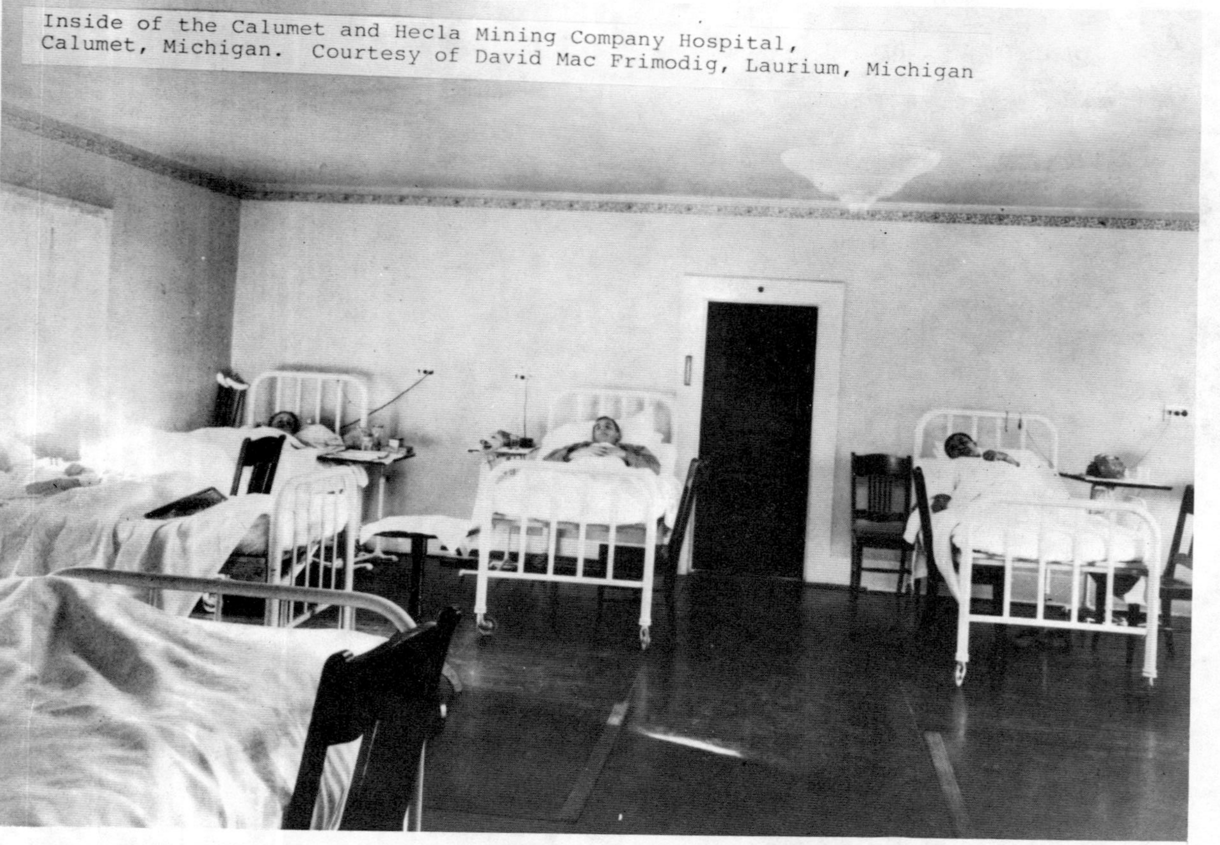

Inside of the Calumet and Hecla Mining Company Hospital,
Calumet, Michigan. Courtesy of David Mac Frimodig, Laurium, Michigan

The Calumet and Hecla Mining Company Hospital
was located on U.S. 41, Calumet, Michigan
Courtesy of David Mac Frimodig, Laurium, Mich

The Calumet and Hecla Mining Company swimming pool in the bath house. Courtesy of David Mac Frimodig, Laurium, Michigan

This picture of the Calumet and Hecla Mining Company bath house was taken in 1913

Underground in the Tamarack Mine, Red Jacket, Michigan

THE CALUMET BREWERY

Calumet Grade School May Festival -1914

The Red Jacket Fire Station was built in 1898 after a fire which almost destroyed the entire village. Courtesy of the Calumet Village Fire Department

One of Calumet's early street scenes

Sinking the Osceola Mine Shaft in 1896

Statements of Condition
Houghton County Banks
May 1st, 1915

"THE OPEN DOOR"

Compliments of the
First National Bank
Calumet, Michigan

For the CALUMET & HECLA CONSOLIDATED COPPER CO., CALUMET, MICH.
By the LAKE SHORE ENGINE WORKS, MARQUETTE, MICH.

Mine timbers for the South Hecla Mine

The Drift & Winse in the Tamarack Mine

Calumet Grade School May Festival -1914

Calumet Grade School May Festival -1914

A C&H copper
mining shaft
with Loie
Conda

ADD THIS COPPER COUNTRY LOCAL HISTORY
SERIES TO YOUR PERSONAL LIBRARY

* *

COR-AGO, A LAKE LINDEN MEDICINE COMPANY
   First of a local history series

A COPPER COUNTRY LOGGER'S TALE
   Second of a local history series

GREGORYVILLE - THE HISTORY OF A HAMLET LOCATED
ACROSS FROM LAKE LINDEN, MICHIGAN
   Third of a local history series

WHITE CITY - THE HISTORY OF AN EARLY COPPER
COUNTRY RECREATIONAL AREA
   Fourth of a local history series

SOME COPPER COUNTRY NAMES AND PLACES
   Fifth of a local history series

THE HISTORY OF LAKE LINDEN, MICHIGAN
   Sixth of a local history series

THE HISTORY OF JACOBSVILLE AND IT'S SANDSTONE
QUARRIES
   Seventh of a local history series

THE HISTORY OF COPPER HARBOR, MICHIGAN
   Eight of a local history series

THE HISTORY OF EAGLE HARBOR, MICHIGAN
   Ninth of a local history series

LAKE LINDEN'S YESTERDAY - A PICTORIAL HISTORY,
VOLUME I
   Tenth of a local history series

THE HISTORY OF EAGLE RIVER, MICHIGAN
   Eleventh of a local history series

JOSEPH BOSCH AND THE BOSCH BREWING COMPANY
   Twelfth of a local history series

COPPER FALLS - JUST A MEMORY
   Thirteenth of a local history series

THE CALUMET THEATRE
   Fourtheenth of a local history series

EARLY DAYS IN MOHAWK, MICHIGAN
   Fifteenth of a local history series

LAKE LINDEN'S YESTERDAY - A PICTORIAL HISTORY,
VOLUME II
   Sixteenth of a local history series

THE KEWEENAW WATERWAY
   Seventeenth of a local history series

A BRIEF HISTORY OF AHMEEK, MICHIGAN
   Eighteenth of a local history series

ALL ABOUT MANDAN, MICHIGAN
   Nineteenth of a local history series

HANCOCK, MICHIGAN, REMEMBERED, VOLUME I
   Twentieth of a local history series

THE SETTLING OF COPPER CITY, MICHIGAN
   Twenty-first of a local history series

LAKE LINDEN'S YESTERDAY - A PICTORIAL HISTORY,
VOLUME III
   Twenty-second of a local history series

PHOENIX, MICHIGAN´S HISTORY
    Thirty-fourth of a local history
    series

FREDA, MICHIGAN, END OF THE ROAD
    Thirty-fifth of a local history
    series

HOUGHTON IN PICTURES
    Thirty-sixth of a local history
    series

THE COPPER RANGE RAILROAD
    Thirty-seventh of a local history
    series

LAC LA BELLE
    Thirty-eight of a local history
    series

TRIMOUNTAIN AND ITS COPPER MINES
    Thirty-ninth of a local history
    series

EARLY RED JACKET AND CALUMET IN
PICTURES, VOL II
    Fortieth of a local history series

A BRIEF LIST OF PUBLICATIONS PERTAINING
TO COPPER COUNTRY HISTORY
    (This publication is not part of the
    series)